In Praise of Mathematics

In Praise of Mathematics

Alain Badiou with
Gilles Haéri

Translated by Susan Spitzer

polity

First published in French as *Éloge des mathématiques* © Flammarion, Paris, 2015

This English edition © Polity Press, 2016

Polity Press
65 Bridge Street
Cambridge CB2 1UR, UK

Polity Press
350 Main Street
Malden, MA 02148, USA

ISBN-13: 978-1-5095-1202-7
ISBN-13: 978-1-5095-1203-4 (pb)

A catalogue record for this book is available from the British Library.
Library of Congress Cataloging-in-Publication Data

Names: Badiou, Alain, author. | Haéri, Gilles, author.
Title: In praise of mathematics / Alain Badiou, Gilles Haéri.
Other titles: Éloge des mathématiques. English
Description: Malden, MA : Polity Press, 2016. | Includes bibliographical references and index.
Identifiers: LCCN 2016012538| ISBN 9781509512027 (hardback : alk. paper) | ISBN 9781509512034 (pbk. : alk. paper)
Subjects: LCSH: Mathematics--Philosophy. | Badiou, Alain--Interviews. | Haéri, Gilles--Interviews.
Classification: LCC QA8.4 .B33713 2016 | DDC 510.1--dc23 LC record available at https://lccn.loc.gov/2016012538

Typeset in 12.5 on 15 pt Adobe Garamond by
Servis Filmsetting Ltd, Stockport, Cheshire
Printed and bound in the UK by Clays Ltd, St Ives PLC

For further information on Polity, visit our website:
politybooks.com

Contents

Contents

Many years ago – a little before and a little after my first philosophical "opus," *Being and Event* (1988) – I introduced the concept of the *conditions* of philosophy, which you'll encounter later in this book. The aim was to identify precisely the real types of creative activity of which humanity is capable and on whose existence philosophy depends. Indeed, it is clear that philosophy was born in Greece because in that country, beginning, at any rate, in the fifth century BCE, there were some totally new ideas about mathematics (deductive geometry and arithmetic), artistic activity (humanized sculpture, painting, dance, music, tragedy, and comedy), politics (the invention of democracy), and the status of the emotions (transference-love, lyric poetry, and so on). So I suggested that philosophy really only develops when new advances emerge in a set of "truths" (that's the name I give them for philosophical reasons) of four different types: science, art, politics, and love. That's why I responded positively to Nicolas Truong's invitation to have a dialogue with him in praise of love, and then in praise of theater, in Avignon. Likewise, I accepted Gilles Haéri's proposal of a dialogue in praise of mathematics in the setting of the Villa Gillet in Lyon. The first two conversations resulted in books published in Flammarion's "Café Voltaire" series. [English translations: *In Praise of Love*, New Press, 2012, and *In Praise of Theatre*, Polity, 2015.] The same is true of the third, which is the subject of this book. All that remains to be done is to write a book in praise of politics, and I'm considering it.

I

Mathematics Must Be Saved

Alain Badiou, you are what I would call, to use a mathematical term, a singularity in the French intellectual landscape.

There's your political commitment, of course, which the general public has been aware of since 2006, with the success of De quoi Sarkozy est-il le nom? [*translated as* The Meaning of Sarkozy, *Verso, 2010*]. *You represent one of the last great figures of the politically-engaged intellectual today, one of the fiercest critics of our liberal democracies, and the tireless defender of the communist Idea, which you refuse to throw out with the bathwater of History.*

But from a more specifically philosophical point of view, the body of work you have produced is also very singular. At a time when philosophy has retreated into specialization, and, in so doing, has renounced its original ambitions, you have consistently attempted to restore meaning to metaphysics by building a system that can be described as a great synthesis on the world and on being. Now, this philosophy, set out mainly in Being and Event, *and later in* Logics of Worlds, *is based to a very large extent on mathematics. You are in this regard one of the rare contemporary philosophers to take mathematics really seriously, and you do not just speak about it as a philosopher but practice it almost on a daily basis.*

Could you begin by telling us where this very strong relationship with mathematics comes from?

It's something that goes back to before I was even born! Simply because my father was a math teacher. So there was the mark of the name of the father, as Lacan would say. Actually, it had a profound effect on me, because I heard mathematics spoken about in my family – by my father and my older brother, by my father and colleagues of his, etc. – in a sort of early imprinting, without my understanding at first what it was all about but sensing that it was at once keenly and obscurely interesting. So much for the first, prenatal stage, so to speak.

Later, as a high-school student, I was fascinated by mathematics as soon as we started doing a few really complex proofs. I must say that what really captivated me was the feeling that, when you do math, it's a bit like following an incredibly twisted, convoluted path through a forest of ideas and concepts, and yet, at a given moment, the path leads to a sort of beautiful clearing. I was struck early on by this quasi-esthetic feeling about mathematics. I think I could mention a few theorems of plane geometry here, in particular theorems of the inexhaustible geometry of the triangle, which we were taught in grades 9 and 10. I'm thinking of Euler's line. First we were shown that the three altitudes of a triangle are concurrent in a point H, which was already great. And then that the three perpendicular bisectors were also concurrent, in a point O – it kept getting better and better! And finally that the three medians were concurrent, too, in a point G! Wonderful. But then the teacher mysteriously told us that it could be proved, as the mathematical genius Euler had done, that these points H, O, and G were moreover all on the same line, which is obviously called "the Euler line"! This alignment of three fundamental points, as the behavior of the characteristics of a triangle, was so unexpected, so elegant! We weren't given the proof, because it was considered too difficult for 10th grade, but our interest in it was piqued. I was thrilled that such a thing could be proved. There's this idea of a real discovery, of an unexpected solution, even if it means you have to make your way along a path that's sometimes a little hard to follow but where you're ultimately rewarded. Later, I often compared mathematics to a walk in the mountains: the approach is long and hard, with lots of twists and turns and steep climbs. You think you're finally there, but there's still one more turn . . . You sweat and strain, but when you reach the summit of the pass, the reward is truly beyond compare: that amazement, that ultimate beauty of mathematics, that hard-won, utterly unique beauty. That's why I continue to promote mathematics from this esthetic perspective, too, noting that it's a very ancient perspective, since Aristotle in fact regarded mathematics as a discipline, not so much of truth as of beauty. He claimed that the greatness of mathematics was esthetic, far more than ontological or metaphysical.

Next, I studied contemporary mathematics in greater depth by taking the first two years of university math. This was from 1956 to 1958, my first two years at the École normale supérieure. I combined significant philosophical discoveries I made there (Hyppolite, Althusser, and Canguilhem were my professors at the time) with the math courses at the Sorbonne and substantive discussions with the math students at the École. It was then, probably also because of the atmosphere of structuralism and the 1960s, when there was a lot of buzz about formal disciplines, that I became really convinced that mathematics was in a very close dialectical relationship with philosophy – at least my conception of it, because mathematics was at the heart of my concerns. Structures are first and foremost the business of mathematicians. At the very end of his seminal book, *The Elementary Structures of Kinship*, the great anthropologist Lévi-Strauss, whom I was reading with passionate interest at the time, referred to the mathematician Weil to show that the exchange of women could be understood by using the algebraic theory of groups. Now, at that time, my philosophical approach required mastering enormous conceptual constructions. What's more, because of its esthetic force and the creativity it calls for, mathematics requires you to become a Subject whose freedom, far from being opposed to discipline, demands it. Indeed, when you work on a mathematical problem, the discovery of the solution – and therefore the creative freedom of the mind – is not some sort of blind wandering but rather the determination of a path that's always lined, as it were, by the obligations of overall consistency and demonstrative rules. You fulfill your desire to find the solution not in spite of the law of reason but thanks to both its prohibitions and its assistance. Now, this is what I had begun to think, first in conjunction with Lacan: desire and the law are not opposites but dialectically identical. And finally, mathematics combines intuition and proof in a unique way, which the philosophical text must also do, as far as possible.

I'll conclude by saying that this back-and-forth movement between philosophy and mathematics produced a sort of split in me . . . and all my work may be nothing but the attempt to overcome this split. This is because my master in philosophy, the one who revealed philosophy to me, was Sartre. I was a convinced Sartrian. But frankly, mathematics and Sartre, as you know, weren't exactly compatible ... He even had a vulgar phrase that he used to trot out all the time when he was young, at the École normale supérieure: "Science is zilch; morality's an asshole."[1] To be sure, he didn't stick to this simplistic maxim, but he never really returned to the sciences, and in particular to the formal sciences. So the conviction grew in me that philosophy should be able to preserve the dimension of the subject, the dimension of the politically committed subject, that sort of historical drama that subjectivity is capable of being, and yet to integrate mathematics in all its rational force and splendor, particularly as regards the doctrine of being.

I could almost sum everything up by saying that it is the overcoming of this split that still constitutes my relationship to mathematics today.

Why do you think it's necessary to praise mathematics today? After all, that discipline is still central to our educational system; it's even one of its primary selection tools. And if one were to judge by the recent French Fields Medalist – bringing to 11 the number

1 Cited in Simone de Beauvoir, *The Prime of Life*, trans. Peter Green (Harmondsworth: Penguin, 1965), 43; translation modified.

of our winners in the field, right after the United States – one might even think that mathematics has pride of place in France. Do you have the contrary feeling that it is under threat?

Well, you know, the vast majority of mathematicians have an extremely elitist relationship to their discipline. They're fine with thinking that they're the only ones who understand it, and that that's just the way it is. After all, they're people who, somewhat out of necessity, essentially speak only to those who are able to understand the most difficult proofs of contemporary mathematics, in other words, their fellow mathematicians for the most part. So we're talking about a very exclusive world, which occasionally attempts to reach out to a somewhat wider public, as does the 2010 Fields Medal winner Cédric Villani, and as did the renowned mathematician Henri Poincaré well before him, but that's still the exception.

So, on the one hand, you've got an inventive, creative mathematics, confined to an extremely close-knit and international, but strongly elitist, world of intellectuals, and, on the other, a sort of school- and university-based dissemination of mathematics, the use of which, in my opinion, has become increasingly unclear and uncertain. This is because mathematics, particularly in France, really *is* used as a method of selection of the elites via the entrance exams for the scientific *grandes écoles*. As the math students used to put it, "We really crammed our asses off for the math exam." But in the end, the organic purpose of all this is still essentially a selective one. This situation has hurt mathematics in terms of its overall relationship to public opinion. The vast majority of people, once they've taken a number of relatively easy exams in school, no longer have any real connection with mathematics. In France, it must be said, it isn't part of ordinary culture. And that, as far as I'm concerned, is scandalous.

Mathematics should absolutely be considered not just as a scholarly discipline tasked with selecting the people who will be engineers or government ministers but as something that's extremely interesting in and of itself. Like fine arts, like cinema, it should be, for reasons I'll come back to, an integral part of our general culture. But, clearly, this is not the case – and it's even less so for cinema, which is perhaps even more scandalous. Because of this, public opinion about mathematics is split between a sort of polite respect for its elitism – bolstered by the usefulness it is credited with having in physics or as regards technology – and an ignorance that can be summed up in the belief "I don't have the math gene." To make a bad pun, you could say that the split is between the very small minority of "gene-iouses" and the vast majority of everyone else. I think this situation is detrimental, even deplorable. But we'll see, perhaps, that it's not so easy to reverse this state of affairs. To put an end to the mathematicians' elitism, a middle way has to be found between the understanding of formalisms and the conceptual aim. And, for that to happen, I think there is a need for philosophy, which should therefore be taught a lot sooner.

You mentioned mathematical applications, which are in fact ubiquitous throughout the contemporary world, even though most people don't understand a whole lot about them or aren't even necessarily aware of them.

There's no question that that's a paradoxical situation: mathematics, today, is everywhere. The new means of communication, which are so fetishized, are based entirely on binary language, new algorithms, prime number coding, and so on and so forth. However, the vast majority of users have no idea what any of it means.

I think this paradox can be clarified by introducing the question of teaching here. What are actually the respective roles, in the development of thought, of knowledge (for example, proficiency in the formal language of mathematics), and the presentation of that knowledge (for example, the real, personal interest that we take in considering the use and implications of these formalisms)? Knowing and thinking, or even loving what we know, aren't the same. They're not immediately identical to each other. What's the relationship between them? This is the key question of transmission. And, as you know, philosophy has always taken an interest in this question. Right from its beginnings. Plato and Aristotle saw themselves as educators. Actually, they regarded philosophy, for the most part, as a didactic, pedagogic enterprise, which may produce new knowledge, of course, but above all sheds light on established knowledge and integrates it into a new subjectivity. This is perfectly the case with mathematics, to which Plato, although dealing with the most advanced knowledge of his time, assigned a general function in the development of thought, of whatever kind. Actually, I'm convinced that philosophy shows us that the question of the transmission of knowledge is relatively homogeneous, regardless of the particular form of knowledge considered. Because, ultimately, the problem of the transmission of knowledge is above all to convince those to whom you're speaking that it's interesting, that they can be enthralled by it. That's the generic problem of all education. You have to convince the people you're speaking to that they have good reason to be interested in mathematics, for example. To be interested in it – as in many other types of knowledge – not for the upward mobility it promises but for itself, for the food for thought it provides. And this is so for anyone you're addressing, without making them think that some people can understand and others can't.

Does this contemporary ignorance of mathematics seem to be the most widely shared commodity in the world,[2] including by your fellow philosophers?

2 Cf. Descartes: "Common sense is the most widely shared commodity in the world."

It's a divided state of affairs. Unfortunately, most philosophers, having minimal mathematics training (often no more than formal logic, moreover), opt for Anglo-Saxon analytic philosophy, or even its scientific satellite, cognitivism. Analytic philosophy focuses on the linguistic distinction between statements that make sense and are reasonable, and those it considers devoid of sense, in particular virtually all philosophical statements since Plato, which are regarded as "metaphysics" and are consequently irrelevant. Cognitivism attempts to reduce all questions of thought or action to the experimental study of brain mechanisms. However interesting the handful of results obtained by these approaches may be, I can't regard them as philosophy. They are academic studies lacking any existential, political, or esthetic interest, which means: unusable for philosophy conceived of as the illumination of real life. Or else, as is often the case in France, mathematical culture encourages people to enroll in an academic "specialization" such as the history of the sciences or epistemology. This also amounts to a renunciation of the true ambitions that ought to animate a philosophical enterprise, which are organized around the question of the meaning of life, of involvement in truths, of what a life worthy of the name can be. Apart from these two – in my opinion! – dead ends, virtually everyone studying philosophy has practically no mathematical culture and thinks that the chief, if not the only, mainstay of their work is the history of philosophy.

The main result of all this is that the real life of mathematics and the real life of philosophy tend to be totally separate from each other. And that's a new situation, at least compared with the more than 2,000 years of philosophy's existence.

Indeed, even though mathematics and philosophy were closely linked very early on – and we'll get back to this later – they are developing differently today.

There's the phenomenon I just mentioned. But there's also what might be called the social or public development of the two groups in question. The contemporary mathematician is usually someone who works on an extremely complex and sophisticated regional area of specialization in mathematics. To be on his or her level, that is, to be able to talk about it with him or her as an equal, is often something, as I said, that less than a dozen people are capable of. Mathematical elitism where creativity is concerned is extremely exclusive; it's the most exclusive of all possible elitisms. Today, given the state of its dissemination, you can't just go into mathematics whenever you feel like. It's not like inherited wealth: it's not passed down, and average or already great, or even very great, knowledge isn't sufficient. As a result, mathematics has become very inaccessible. Strictly external references exist and are reported in the press: someone who has discovered something very important will win the Fields Medal, with the approval of his tiny community, and, moreover, amid a widespread lack of comprehension.

When it comes to philosophy, the problem is the exact opposite, since just about anyone can be considered a philosopher now. Ever since philosophers have become "new," people are very undemanding where they're concerned, even at a basic level, I can assure you! In Plato's, Descartes', and Hegel's time, or even in the late nineteenth century, the knowledge requirement for claiming you were a "philosopher" concerned virtually all the different types of knowledge and the political, scientific, and esthetic discoveries of the time, while, today, all you need to have are opinions and the right connections in the media to make people think those opinions are universal, whereas they're totally banal. Yet the difference between universality and banality should be crucial, after all, for a philosopher.

It is alleged that it has become impossible to have such extensive knowledge today. But that's not so. Naturally, we can't master the whole extent of the field of the sciences, or the whole of the world's artistic production, or all political inventions without exception. But we can, and we must, know enough about them, have a sufficiently deep and broad experience of them, to be able to legislate philosophically. However, many "philosophers" today fall far short of this minimal standard, especially when it comes to the science that has always been the most important one for philosophy, namely, mathematics.

This is a fairly recent state of affairs, inasmuch as it only developed in the late 1970s and early 1980s. It has seriously damaged the image, the idea, the conception, of the philosopher. A philosopher has become merely a consultant in anything and everything. I myself, I must admit, am exposed to this corrupting temptation. When

I wrote *Ethics* in the early 1980s, I received a lot of invitations to give banking ethics seminars. I'm saying this seriously – I can produce the documents! These people couldn't have cared less about either my opinions or my commitments: since I had talked about ethics, they thought it was only normal for me to be in the service of what they regard as the heart, the living center, of society – the bank!

So the divergence between mathematics and philosophy also stems from the fact that philosophy, owing to the shallow, reactionary figure of "the new philosopher," has undergone an incredible trivialization of its status. The philosophy media stars are, it must be said, and strictly in terms of the knowledge required to talk about what they talk about, idiots. In mathematics, they'd be considered the equivalent of a very average high-school senior. This is, incidentally, an important virtue of mathematics: it's impossible to have frauds of that sort in it. But the flip side of that virtue is that mathematics has become out of reach, or the object of bitter indifference, because of its elitist isolation from the other regimes of knowledge. Obviously, with such a rigorous selection process, we haven't had any "new mathematicians," that's for sure. And I don't see how there could be any. A "new mathematician," even today, is someone who proves – either with great difficulty or brilliantly – previously unknown theorems, and you can't produce imitations or fakes of those, it's absolutely impossible.

So we're dealing with a degree of separation between mathematics and philosophy that would have astonished most of our great classical or modern ancestors, many of whom, and some of the most famous ones, I should point out, were also great mathematicians. Descartes was a foundational mathematician, the inventor of analytic geometry, which is a sort of unification of geometry and algebra: he showed how a curve in space, hence a geometric object, can be represented by an equation. Leibniz was a mathematical genius, the founder of modern differential and integral calculus. The last ones who even came close to them lived sometime in the nineteenth century: perhaps Frege, perhaps Dedekind, perhaps Cantor in some respects, or Poincaré, who was certainly the last great figure of that particular model. There was also a philosophical school in France, between 1920 and the 1960s, that was proficient in mathematics and yet did not succumb

to the siren song of so-called analytic philosophy. Its members included Bachelard, Cavaillès,
Lautman, and Desanti. But today, the separation
is very advanced, even though twenty or thirty
years after me there has arisen a new generation
of philosophers, and of a few mathematicians, too
(Tristan Garcia, Quentin Meillassoux, Patrice
Maniglier, et al.), a very promising generation,
generally speaking, thanks to its rediscovery
of metaphysics. Some of them have mastered
a significant part of the field of contemporary
mathematics without immediately reducing it to
a sort of linguistic positivism or a mere history of
the sciences. I'm thinking in particular of Charles
Alunni, René Guitart, Yves André, and, more
recently, Elie During and David Rabouin. I'm
obviously forgetting – or else I don't know about,
which I hope is the case – many other talented
people in the upcoming generations.

In fact, part of my efforts specifically related to
metaphysics are devoted to trying to overcome
– with the help of everyone who has the means
and the desire to do so today – this deadly separation between everything that goes by the name
of philosophy and the tremendous intellectual
discoveries of contemporary mathematics.

II

Philosophy and Mathematics, or the Story of an Old Couple

I'd like for us to explore in greater detail the links between philosophy and mathematics. You mentioned a moment ago that they were an old couple. Plato had already inscribed over the entrance to his Academy: "Let no one ignorant of geometry enter here." How would you account for this close association?

Mathematics and philosophy have indeed been connected right from their beginnings, even to the point where a variety of particularly famous philosophers – Plato, but also Descartes, Spinoza, Kant, and Searle – categorically declared that without mathematics there would have been no philosophy. So mathematics was conceived of very early on – and entirely explicitly in Plato's case – as a sort of precondition in order for rational philosophy to come into being. Why? Simply because mathematics exemplified a knowledge process that "held up on its own," so to speak. In other words, when you've got a proof, well, you've got a proof! This is nothing like when truth is proclaimed by a priest, a king, or a god. The priest, the king, or the god is right because they're a priest, a king, or a god. What's more, if you disagree with them, they'll let you know about it . . . Whereas for mathematicians it's completely different: they have to construct a knowledge process that will be shown to their colleagues and rivals. And if their proof is false, they'll be told so.

So from very early on, from the time of ancient Greece, mathematics was a world in which things considered to be true, to be proven, could circulate provided they were validated and accepted by the community of people who were "knowledgeable about it," and not just because of the authority stemming from the mathematician's calling himself a "mathematician." On the contrary, the mathematician was somebody who, for the first time, introduced a universality completely free of any mythological or religious assumptions and that no longer took the form of a narrative but of a proof. Truth based on a narrative is "traditional" truth, of a mythological or revealed type. Mathematics swept aside all the traditional narratives: the proof depended only on a rational demonstration, shown to everyone and refutable in its very principle, so that someone who had put forward a hypothesis that was ultimately proved to be false had to accept that he was wrong. In that sense, mathematics is part of democratic thought, which moreover appeared in Greece at the same time. And philosophy could only be constituted in its – always threatened – autonomy from the religious narrative thanks to this formal support, which no doubt concerned a limited area of intellectual activity but one that had totally independent norms, explicit norms, which everyone could know. A proof had to be a proof, and that was all. So it's true that, from the very outset, there were close links between mathematics, democracy (in the sense of political modernity, as opposed to the traditional authorities), and philosophy.

So, in historical terms, mathematics originated before philosophy?

It's a complex and poorly documented story. I share the historian and philosopher of science Arpad Szabo's view: if you look closely at the thinking of Parmenides or the whole "Eleatic" school (so-called because it was made up of the citizens of Elea), prior to Socrates and Plato, hence dating back to the fifth century BCE, you can see the deep trace of methods of thought that would reach their full realization in mathematics. This is the case with *reductio ad absurdum*, which I consider to be decisive in the intellectual power created by the mathematics of that time. I explored this issue in detail in my 1985–6 seminar devoted to Parmenides. Roughly speaking, *reductio ad absurdum* amounts to proving that a proposition p is true not by directly "constructing" its truth from already established truths but by demonstrating that its opposite proposition, i.e., the proposition not-p, is necessarily false. You then apply the principle of the excluded middle: "Given p, a well-formed proposition (one that obeys the syntactic rules of the system in question), either p is true or not-p is true.

There is no third possibility." This is a remarkable process because it proves a truth by operating entirely within a false hypothesis. Indeed, how can it be proved that not-p is false? Simply by assuming that it is true and by deriving from this hypothesis consequences that contradict already established truths. You then apply the principle of non-contradiction: since not-p contradicts a proposition – let's say q – that is true, and since two contradictory propositions can't both be true, not-p has to be false. And therefore p has to be true.

You can see the amazing path of the proof. You want to prove that p is true, and you have your reasons for this (it's your hypothesis). To that end, you fabricate the fiction "not-p is true," which you hope is false! And to feed your hope, you draw consequences from this fiction, thus operating with implacable logic within what you think is false, until you come up against a consequence that explicitly contradicts a proposition that was previously proved to be true. This controlled, regulated navigation between the true and the false is, to my mind, completely characteristic of nascent mathematics, of the break it introduces with respect to any revealed truth or truth whose force is only poetic. Now, this is a "tone" we find in Parmenides. And the reason we do is that, in order to prove that being is, that this is the fundamental truth, he first proves that not-being is not. He therefore uses reasoning by the absurd. My conclusion is categorical: rational philosophy and mathematics originated at the same time, nor could it have been otherwise.

You pointed out that, subsequent to the Greeks, the classical philosophers always took a very close interest in mathematics. Did it really have an influence on their systems of thought?

It's interesting to consider the reasons the philosophers themselves gave to explain the importance of mathematics.

Let's consider Descartes, the founder of modern philosophy. As I pointed out, he was a very great mathematician. What he took from mathematics in terms of his specifically philosophical project is clear: it was the ideal of the proof. For him, the philosophical text had to take the form of those "long chains of reasoning" that are the essence of mathematics. But it could also be said that he used the detour through the absurd. Indeed, to prove the existence of something, the existence of the outside world, he didn't proceed directly but instead invented the fiction of an "absolute doubt," a "hyperbolic" doubt, which would amount to asserting the nothingness of all truth and experience. And he then observed that the very fact of doubting could not itself be doubted. This is the famous "cogito" argument, which established a "point" of truth (the "I exist") through negation of the absolute negation that is doubt. Furthermore, to prove the existence of God, Descartes would explicitly provide several different proofs that were, generally speaking, positive ones this time. For instance, from the fact that it is certain that I have an idea of the infinite, whereas I am finite, it follows that there must be an infinite being that created this idea in me. The details of the proof are more complex, more "mathematical," in a word . . . With Descartes, mathematics is ubiquitous, as the paradigm of rational thought.

Let's take Spinoza, still in the seventeenth century. He began his *Ethics* by saying that if mathematics hadn't existed, man would have remained in ignorance, in particular because he would have continued to explain everything by "final causes," mythologies, or the influence of supernatural powers. Spinoza himself thus inscribed his ethics in the idea that it was in a certain sense a possible consequence of the existence of mathematics. The key role of mathematics, in his view, was to have discredited explanations by final causes, to have expelled from the philosophical field finality, which was still so important in the Aristotelian tradition, and to stick to deductive reasonings. Spinoza distinguished three types of knowledge, as indeed Plato had done. The first is a combination of sensory and imaginative representation, or what could be called ordinary ignorance. The second is methodical conceptual knowledge, the step-by-step proof, and mathematics is the paradigm of this. The third type is the intuitive knowledge of God, who is the name of Nature or the All, and this is specifically philosophical knowledge. But Spinoza made it clear that without access to the second type, reaching the third type would be out of the question. What's more, he organized his book in the exact same way as the mathematical treatises of his time were organized, on the model of Euclid's *Elements*: definitions, postulates, propositions, etc. Philosophy was thus set out *more geometrico*, in the geometric mode. That goes to show how close a connection there is between philosophy and mathematics.

A hundred years later, what did Kant say about mathematics? In the introduction to his *Critique of Pure Reason*, he repeated how absolutely necessary mathematics was in order for philosophy to exist, especially critical philosophy, which, in the spirit of the Enlightenment, he intended to found. There would have been no point to the critical question he raised – "Where does the universality of the sciences come from?" – if there had been no science, nor would there have been, as Newton is the proof, any natural science if there had been no mathematics. He also added, and this has always touched me, that the invention of mathematics resulted from "the happy inspiration of a single man,"[3] who, in his mind, was Thales. So Kant also wanted to show that the emergence of mathematics was not a historical necessity but a creative contingency. Mathematics was not created so that Kant could ask the critical question of where rational univer-

3 Immanuel Kant, "Second Preface," *Critique of Pure Reason*, ed. and tr. Paul Guyer and Allen W. Wood (Cambridge: Cambridge University Press, 2000), 106.

sality came from; it was created by chance, one day, from the happy inspiration of a single man, as though it was a kind of serendipitous esthetics. But this serendipity created the possibility of the critical question, which defines the philosophical enterprise.

But another point still needs to be added, which anticipates the interplay between the two possible conceptions of mathematics I'll discuss in a moment, conceptions that have been competing with each other for hundreds of years: the realist (or Platonic) conception, which holds that the object of mathematics exists outside of us, and the formalist conception, which holds that mathematics is a pure creation, and in particular the creation of a special formal language. Kant's conception of mathematics is an "a prioric" conception, meaning that the organization of mathematical thinking does not originate in concrete experience but is prior to it; it exists, with regard to experience, *a priori* and not *a posteriori*. In a nutshell, Kant claims that what is at stake in the formal sciences – and also, though this is a different question, in the experimental sciences – is the subjective organization of human knowledge, of what he calls "the transcendental subject." If rationality is universal, in Kant's view it is not so because it touches a real but rather because it refers to a universal structure of cognitive subjectivity itself. If everyone is in agreement about a mathematical proof, it's not because it refers to anything that touches the thing in itself or the real of the world; it's because the structure of the human mind obeys a single paradigm, such that what will be a proof for one person will be a proof for another. I think this is a sophisticated version of the formalist thesis. Later, for Wittgenstein, mathematics would only be one language game among others, which should not be absolutized. Kant wouldn't say as much, since he considered mathematics to be really universal and irrefutable for minds like ours. But it's a formalism nonetheless, a transcendental formalism: mathematics isn't universal because it thinks formal structures of being *qua* being but because it is a language that's coded in the same way for everyone. However, for Kant, as for Descartes and Spinoza, mathematics, once invented by Thales, paved the infinite way for science, and if it didn't exist – after all, human beings existed for tens of thousands of years before the Greeks invented demonstrative geometry and arithmetic

– the philosophical question (where do universally true judgments come from?) couldn't have been formulated or answered.

You seem to be suggesting a sort of priority of mathematics over philosophy.

There are only two approaches when it comes to this issue, only one of which, as far as I'm concerned, is valid. I think the basic relationship between philosophy and mathematics is actually a reverential relationship, so to speak. There is something about philosophy that defers to mathematics. If indeed philosophy does not defer to mathematics, then it neglects it or rejects it; it thinks, as does Wittgenstein, that there is nothing in mathematics that concerns human existence – this is the second approach I was talking about, which I completely disagree with. There are no half-measures. To be sure, we know full well that "new philosophers" are utterly uninterested in mathematics. They're interested in public opinion, in the Muslim religion, in "totalitarianism," in the cantonal elections, in lots of things, but not in mathematics. And in my view that's an offense. It's an offense against the imperative of rationality that was slowly worked out and established by the great history of philosophy, regardless of the ultimate conclusions, assertions, and positions of the various philosophers. Between Plato's passion for mathematics and Hegel's harsh critique of the strictly mathematical concept of infinity there is a huge gulf. But Hegel was knowledgeable about the mathematics of his times, namely, the work of Euler. In his *Logic*, he devoted a perspicacious note to differential calculus. I don't have anything against the various assessments of the importance of mathematics, but I do have something against the indifference to it and the ignorance, which in my opinion are such serious offenses that they preclude anyone from calling him- or herself a philosopher, even with the epithet "new" attached to the word. And I even went so far as to speak about a reverential relationship, because philosophy cannot just run into mathematics by accident or as though it were just some ordinary topic of epistemology. It can only be seized by mathematics at its very beginning. As the science of being, mathematics is crucial right from the start, as soon as one gets into philosophy. I agree wholeheartedly with the maxim of Plato's academy, which I am repeating on my own behalf: "Let no one ignorant of geometry enter here." And "here" is not just an academy; it's philosophy itself.

Another important issue is the fact that, to a great extent, mathematics escapes the singularity of languages. Naturally, when you teach mathematics in China, you speak Chinese, but, ultimately, mathematics in and of itself belongs to no one language. There is a sort of mathematical language, but it's not French or English or Chinese. In a way, this language, which can always be formalized and reduced to a series of signs in accordance with fixed rules, is beyond-language. But philosophy has always been concerned about the problem of the multiplicity of languages, since it can always wonder: "What does my thinking owe to this language that is particular? Doesn't the particularity of a language make my supposedly universal discourse less universal than it aspires to be?" And it is well known that there are even a few philosophers who were tempted to say: "Yes, but certain languages have universal significance." Some suggested German while others – often the same ones – suggested Greek. It is absolutely remarkable that Descartes should be one of the rare philosophers to say that this question was of no interest to him and to explicitly claim that Reason can be understood in the same way in any language, even, he said, in "low Breton." But this question of languages is a problem, like it or not. Mathematics, however, is a thought process that bypasses the particularity of language. Why? Because one's native language, one's everyday language, is not, strictly speaking, the language of mathematics. It is the language used in explaining it, or in learning it, which is not the same thing.

But don't go thinking I think philosophy should admire, and even revere, mathematical language exclusively. Not at all! Mathematics is concerned with, or latches onto, the most formal, abstract, universally quasi-empty dimension of being as such. It's easy to claim, as we'll see later on, that everything that exists forms a multiplicity. So I'll argue that, since mathematics is the general theory of the different forms in which multiplicities acquire a certain consistency, it is a theory of that which is, not insofar as it is this or that, but simply insofar as it is. Yet, the relationship of thought to being *qua* being is certainly not the whole of subjects' relationship to the world, absolutely not. Mathematics is not the science of the difference between autumn foliage and a summer sky; all it says is that, in any case, all of that is multiplicities, forms that have something in common: the fact of being, quite simply. And it is the abstract forms of this "common" that mathematics attempts to think.

This is a philosophically necessary, but certainly not sufficient, experiment. I for my part use poetry at least as much. Poetry is the other extreme of language, because poetry is what delves into language so as to force it to name what it couldn't name before. And so poetry burrows into the native language, into the particularity of a given language. But within this particularity of the language it will engage in description, transposition, metaphorical comparison, and so on, to such an extent that, in the end, it, too, will touch something universal. It could be said that the poem amplifies the particularity of the language to its limit, to the point of beyond-language, while mathematics from the outset operates outside the particularity of languages. Two contrasting paths but both leading to the real, to universality.

But are the mathematics that are being developed in India, in France, or in China today all the same? Are they really impervious to cultural or linguistic specificities? If so, that would confirm the admirable universality that you were talking about.

Ultimately, yes. If there's a genuine Internationale, today, it is really that of mathematicians. They no doubt speak English, as everybody does, among themselves, but above all they "speak mathematics" – as in fact we all should be able to speak "communist politics" someday, even if in English . . . There are of course schools of mathematics, "historical moments" in mathematics, with national overtones. Let's not forget that in the Middle Ages Baghdad was the indisputable capital of mathematical thinking. And I can give a few other random examples. At the time of the French Revolution or Napoleon, a brilliant French school of geometry grew up around Monge. In the mid-nineteenth century, Germany shone its brightest, with Riemann, Dedekind, and Cantor. In the 1920s and 1930s, the Polish school of mathematical logic, featuring Tarski in particular, was altogether remarkable. In the wake of the truly extraordinary Ramanujan, we can speak even today of an amazing Indian school of number theory. In that area, moreover, the English, from Hardy to Wiles, have not lagged behind. Many other Russian, Italian, American, Brazilian, Hungarian, etc. examples could be cited. It's clear that mathematics has gradually brought founding geniuses to light in practically every region of the world. But, every time it has done so, their work has been enthusiastically adopted by the worldwide fraternity of mathematicians, without issues of language and culture coming into the picture in any significant way. Thus, we could say, yes, mathematics obviously and inexorably cuts across national particularities, without ever getting caught in them, as all truth procedures, including the seemingly most "cultural" ones, such as the arts and, of course, politics, should, and will, do. This is an additional reason why philosophy, which has created universality as its own value, should revere the Mathematicians' Internationale.

We may nevertheless have the impression today that this dialogue between mathematics and philosophy, or this reverence you were talking about, has been doubly shattered: you noted the fact that philosophers have little interest in mathematics but, by the same token, many leading scientists, physicists, and mathematicians practice their discipline uncritically. As though a sort of positivism had taken hold, allowing people to do mathematics or the sciences without reflecting on their universality, their own particular truth. How do you explain this?

It's the philosophers' fault. Frankly, I absolve the mathematicians! There are surely some philosophers among them: as I said, in the past, from Descartes to Poincaré, that was an established fact, but it's still the case today. In the area of mathematics I know best, modern set theory, I can say, for example, that Woodin's meditation on the different meanings of the word "infinite" – Woodin being without doubt the most impressive specialist of what's called "descriptive set theory," namely, the fine structure theory of real numbers – has an undeniable philosophical quality to it. That said, mathematicians have always been entitled to do mathematics day and night for their own personal satisfaction, or for the satisfaction of showing off to the handful of fellow mathematicians who understand the same thing as they do. So they can delve deep into a difficult problem without asking themselves every time whether mathematics is an ontology or a language game. I forgive them their shared negligence of philosophy, because by devoting their lives to such an arduous, seemingly thankless, or grueling pursuit, they are rendering an invaluable service to humanity as a whole.

Besides, we have to face facts: there are plenty of mathematicians who are weirdos, tortured or strange personalities. Take, for example, Grigory Perelman, that absolutely brilliant contemporary Russian mathematician who proved a 100-year-old conjecture that had resisted the efforts of a host of leading experts. Well, he lives as a hermit in a cabin in the woods, is cut off for the most part from the outside world, talks only to his elderly mother, refused the Fields Medal, the honor coveted by the whole mathematical community, and so on. He's a mystic, actually, and he is in that sense a sort of spiritualist philosopher, in the Russian tradition. The two greatest founding geniuses of set theory and mathematized logic, Cantor and Gödel, were both very strange. The former wrote to the Pope to verify the orthodoxy of his thinking of the infinite, then invented a new theory according to which Shakespeare wasn't

Shakespeare. The latter was afraid that some of his colleagues were poisoning his tap water. Just look at a young genius like Évariste Galois, who invented the algebraic theory of groups and, more generally, the constructive spirit of modern algebra. He was a typically Romantic character, who, when arrested for rebellion in the spirit of "The Three Glorious Days" of 1830, wrote down his amazing thoughts day and night in prison and died in 1832, at the age of 20, in a stupid duel over a girl who, as he wrote to his best friend just before getting himself killed, wasn't really worth it. Sure, there were also towering geniuses, like Gauss and Poincaré, who were serious academics, thoughtful people who were well established in their social world. But mathematicians, like poets, can also be anarchistic and romantic, or contemplative and withdrawn, people, because what ultimately matters in mathematics is inventiveness, which often comes to them, after long nights of slow and uncertain work, in the form of a sort of lucky intuition. There's a famous text in which Poincaré explains that a problem he'd been sweating over for weeks and weeks suddenly became clear to him as he was putting his foot onto the step of a bus. That's what mathematics is about, too. So let's not give it a hard time. There are no "new mathematicians" whose only desire is to bolster the dominant reactionary politics – that's something at least.

So it's the philosophers' fault if philosophy and mathematics have parted company?

Absolutely. And not just because of their partial deterioration but because, from a certain point on, philosophers – for pretexts and reasons that should be examined – gave up thinking that philosophy could assume all of what I call its conditions, which I reduce to four "types," these being for me different kinds of what I call truths: the sciences, cognitive truths; the arts, sensible truths; politics, collective truths; and love, existential truths. Most professional philosophers of our time have given up thinking that philosophy – as it clearly claimed to do in Hegel's time, or still later in the time of Auguste Comte, Searle, or Bachelard – requires, and this is a strict minimum, as extensive a relationship as possible with this very complex system of conditions. Our professional philosophers have given up thinking that the idea of a specialized philosophy actually made no sense. That philosophy might be the philosophy of this or that, that it might have special "objects," is what Lacan called the "discourse of the university," in the worst sense of the term. Philosophy is philosophy, or, in other words, something that entertains a special and comprehensive relationship with the sciences, the arts, politics, and love. So there has been a serious capitulation on the part of philosophers.

When did this capitulation, this "separation" between mathematics and philosophy, occur in historical terms?

In my opinion, there was a turning point that began in the late nineteenth century, a turning point that I would term anti-philosophical in a certain way, with brilliant personalities like Nietzsche or Wittgenstein, big stars whose genius I acknowledge but who moved philosophy's agenda in a direction that had not been its direction since Plato. In particular, it was they who abandoned the idea that the *comprehensive* and systematic nature of philosophy had to be accepted, and this resulted in the risk of an indifference to mathematics. In my view, this rupture is especially serious in that the mathematics in use from the late nineteenth century on was in fact mathematics that drastically changed many things in the most essential philosophical concepts.

Could you give us an example?

I'll focus on the concept of infinity, its history, and the contemporary state of the question and its consequences. On this issue alone, breathtakingly new and important research has been carried out in mathematics over the past fifty years. If you're not familiar with it, what happens is that, when you say the word "infinity," you actually have no idea what you're talking about, because the mathematicians have worked on this concept and taken it to an unprecedented degree of complexity. If you don't know anything about certain theorems from the 1970s or 1980s on the new figures of mathematical infinity, there's no point in using the word "infinity" – at least in the context of rational thought.

Likewise, in philosophy "logic" continues to be spoken about, but if you don't take a close look at what has been going on in logic in terms of its constant formal re-creation, you'll have a poor and false understanding of the word "logic." In fact, logic, or rather logics, have become part of mathematics today. I'll come back to this. But it's clear that philosophers cannot be unaware of logic, and therefore of mathematized logic today.

These two examples show that philosophy, if it separates from mathematics, heads for disaster, since a considerable number of the concepts it needs become, simply as a result of ignorance, obsolete.

To sum up, I'd say that there's been a break between mathematics and philosophy. There are historical reasons for it. Philosophical romanticism, from Hegel to Sartrian existentialism, moved away from analytical and demonstrative rationality. And, beginning with the French Revolution, the new concern for history valued movements, revolutions, and negativity to the detriment of the kind of *sub specie aeternitatis* contemplation of mathematical truths, which become timeless once they've been proved. There were also institutional reasons: the growing academic separation between disciplines, the organization of literary and scientific studies into two strictly separate entities. Whatever the case may be, this break has had disastrous consequences in terms of philosophy itself. It has led to the abandonment of the real conditions of the existence and formation of concepts that are still used in philosophy, with the philosophers lagging several miles behind what the mathematicians have defined and proved concerning these concepts.

I fear that it will take quite a while to remedy this situation, but we need to begin promoting the pleasure of mathematics, on the one hand, and restoring the ambition of a rational metaphysics, on the other.

III

What is Mathematics About?

Before we go any further, I think it's important to define mathematics a little more precisely. Russell said it was the field "where we don't know what we are talking about, nor whether or not what we say is true." Could you nevertheless say a bit more about it?

Good old Russell! To begin with, I'd like to point out that the question of the definition of mathematics isn't a mathematical question. That's a very important point. As soon as you get into the question of "What is mathematics?" you're switching over to philosophy, you're doing philosophy. Philosophers have taken a great interest in this question and have even gotten some mathematicians interested in it – the ones with the broadest encyclopedic culture, people like Poincaré, or even Grothendieck more recently – but it remains a philosophical question nonetheless.

We can obviously begin with a sort of basic description. Starting with the Greeks, mathematics has dealt with several related areas. For the Greeks, there were essentially two such areas. First, geometry, which studies objects and structures in space: in two dimensions, plane geometry (triangles, circles, etc.) or in three dimensions, geometry in space properly speaking (cubes, spheres, etc.). Second, arithmetic, which studies numbers. The link between the two is the very important and difficult question of measure: a line segment, once a unit of measurement has been determined, possesses a length, which is in fact a number. This is why there immediately arose very complex problems that right from the beginning of demonstrative mathematics produced a sort of combination of geometry and mathematics. A famous example: if you know the length of the radius of a circle, can you find the circumference? It is here that the number π appears: if r is the length of the radius of the circle, then the circumference is $2\pi r$. The remarkable thing is that the real nature of the number π would only be established in the nineteenth century: only then would it be proved (and it wasn't easy!) why π can't be a whole number, or the ratio of two whole numbers (a fraction, which is also called a rational number), or even the solution of an equation whose coefficients are whole numbers. These numbers that resist simple calculation are now called "transcendental" numbers and by themselves constitute a significant part of modern mathematics.

This fundamental distinction between "spatial" structures and "numerical" structures remains today, in a much more highly developed form. The great "comprehensive" treatise of modern mathematics, undertaken in France in the 1930s by a group of mathematicians who gave themselves the name "Bourbaki," makes a distinction from the outset between algebraic structures, which are the possible structures (addition, subtraction, division, root extraction, etc.) that enable calculations, and topological structures, which make it possible to think spatial arrangements (neighborhoods, inside and outside, connections, the open and the closed, etc.). This is obviously descended from the distinction between arithmetic and geometry. So the most complex and exciting mathematical problems are clearly those that combine the two orientations, specifically the daunting problems of algebraic geometry.

But we're only at an elementary descriptive level here. The real philosophical problem is to define the nature of mathematical thinking in general, whatever its area of inquiry. Now, as far as this issue is concerned, there have historically been some answers that seem to vary widely. However, I think, as I said a moment ago, that there are two main orientations. First, the one that aligns mathematics with an ontological, or, at the very least, "realist," shall we say, vocation, which mathematicians themselves often call "Platonic." In this view of things, mathematics is part of the thinking of what there is, of what is. As to in what respect, how, and so on, it's quite complicated. But let's just say at this stage that mathematics is a way of approaching the real, including the most elusive real. And this is basically because the assumption has to be made that there is an aspect of generality or universality to what exists that is somehow immaterial. There are structures that recur in everything that exists.

The study of these structures as such, of structural possibilities, is precisely the aim of mathematics.

This moreover explains something very strange – which even Einstein was amazed by – namely that physics, i.e., the scientific theory of the real world, couldn't exist without mathematics. As Galileo, one of the founders of physics, essentially said, the world is written in a mathematical language. This first orientation maintains that mathematics has an essential relationship with everything that exists.

Then there's another orientation, which I call "formalist" and which amounts to saying that mathematics is merely a language game, or, in other words, the codification of a language that is of course formally rigorous, since the concepts of deduction and proof are normative and formalized in it, but whose rigor cannot claim to have an ongoing relationship with empirical reality. The oft-cited argument in support of this theory is: "Mathematical axioms can change, after all" and so there is more than one possible mathematical universe. This debate began in the early nineteenth century, when it was understood that there was more than one kind of geometry: Euclidean geometry, which had reigned supreme until then, but also Lobachevskian and, later, Riemannian, geometry. Let's review that history. For centuries it was taken as self-evident that, through a point outside a line, there can pass one and only one line parallel to the given line. This obvious fact was dictated by our perception. Time and again, attempts to prove it from the other axioms of classical geometry all failed. But then Lobachevsky (1829) rejected the axiom, stating that more than one line parallel to a given line can pass through a point outside the line. And rather than ending up with a contradiction, he thereby invented a geometry that was non-Euclidean but consistent and productive. Later, Riemann (1854) assumed the axiom that there exists no line parallel to a given line. And this was not just consistent with the other classical axioms but gave Einstein and relativistic physics a natural geometric framework. And then today, when mathematical structures of all sorts abound, we may have the feeling that it's all a sort of free-wheeling human creativity, which gives itself first principles, axioms, and specific logical rules and derives the consequences from them, but that it's ultimately just a formal game. A remarkable mental game in which the demonstrative processes have to be identified as rules – the rules of the game – and the axioms as the initial data of the game. And the consequences are what you get when you apply the rules to the initial data. So a great theorem is nothing but a well-played game, a game won. As we know, this is the path taken by the anti-philosophical logician Ludwig Wittgenstein, with all the brilliance he was capable of. It is in my view very symptomatic that his having regarded mathematics as a pure language game, ultimately without any real seriousness, led him to a sort of ironic contempt for the highest ambitions of contemporary mathematics. He heaped sarcasm, for example, on set theory. The fact is, one of the greatest creative minds, in pure logic as well as in set theory, namely, Kurt Gödel, was a convinced Platonist. All throughout the last century the conflict between the realist and formalist – or linguistic – orientations was so fierce that indisputable geniuses, philosophers and/or mathematicians, could find themselves on opposite sides. This debate about what mathematics is has actually existed right from the start, however. I mentioned that Aristotle regarded mathematics as esthetic above all. He therefore viewed it as unrelated to the real, as an arbitrary creation that produced a certain pleasure of thought. For Plato, on the contrary, mathematics was the very foundation of universal rational knowledge: the philosopher absolutely had to begin with mathematics. Even if he ultimately went beyond it, he had to learn mathematics first. Plato thought that political leaders, for example, would be well advised to study higher mathematics for at least ten years. He indicated that they were not to be satisfied with just the minimum, since they had to do geometry in space in particular. As geometry in space had only just emerged in Plato's time, it could be said that, for Plato, the true leader of the ideal state had to be like the mathematical genius Henri Poincaré, not like the very reactionary president Raymond Poincaré, who was in large part responsible for World War I. Basically, for Plato, the right method would have been to choose Nobel Prize or Fields Medal winners as presidents of the Republic. It's clear that this is a completely different political alternative from the one prevailing today . . .

In the formalist conception of mathematics, the initial axioms have a status of being arbitrary, independent of our intuition, or, in other words, having no absolute truth value. But isn't that actually pretty bogus? Can anyone really think that an arbitrary definition would create a mathematical object, such as natural numbers, for example? Isn't it rather because the natural numbers pre-exist and have necessary properties that we can then try to express or formalize them by axioms? Like when Russell, for example, reconstructed the concept of number on the basis of set theory: all triplets, sets containing three elements, form a family of sets with which the number 3 will be associated. Fine, but can you really speak of triplets without already having an intuition of the number 3? Isn't that a strange sleight of hand?

Yes, no doubt . . . But you see, the intuition of the number 3, which has probably been accessible to the human animal since its origins, doesn't in and of itself produce any mathematics yet. If, on the other hand, you write the number 235,678,981, it doesn't correspond to any kind of intuition. It doesn't represent anything to you that you can intuitively distinguish from 235,678,982. Except by writing, but the writing of what? That's the whole question. Mathematical thinking makes a tentative appearance if you say that 235,678,982 is the "successor" of the number 235,678,981.

But you can then see that what really matters is the word "successor," which actually denotes an operation and therefore, ultimately, a structure, in this case addition: if the number n exists, whatever n may be, then there also exists the number $n + 1$, which will be called the successor of n. But why *the* successor? Couldn't there be more than one of them? No, that's not possible, because the additive structure of natural numbers requires that no other number exist between n and $n + 1$. But then you'll say, what does "between" mean? Well, the word refers to another structure, the order structure, which formalizes – and profoundly transforms – the concepts of "greater" and "smaller." If n is smaller than q and q is smaller than r, then q is "between" n and r. The notation we're all familiar with illustrates this in an almost spatial way: we write $n < q < r$. All this amounts to saying that the natural numbers at any rate have the algebraic additive structure and an order structure. It will then be noted that this order structure is "discrete" in the following way: there are "holes" or "gaps" in the chain of order. Indeed, there is no natural number between n and $n + 1$. If we only take natural numbers into account, we can really say that there is nothing between n and $n + 1$. This "nothing," if we say that it's a number (which the Arab algebraists were the first to dare to do), will be integrated, under the name of "zero," into the additive structure in the following way: if you add zero to a number n, well, you still have n. Zero is said to be the neutral element for addition. And it will also be integrated into the order structure, in that zero, as the name of nothing, is surely smaller than all the other numbers. It is therefore a minimum for the order structure.

You can continue going through the natural numbers like this in the articulation between many structures: addition, multiplication, division, prime factor decomposition, and many more. You will then have established, far from the primordial, infra-mathematical intuition of the 1, the 2, or the 3, a magnificent science: basic mathematics. It's a great temptation, in this case, to say that natural numbers are reducible to a structural web, itself the result of axioms that can be changed in order to obtain the formal essence of other so-called intuitions. Let me give you just one example: I said that there exists no number between a number n and $n + 1$ when we clearly have $n < n + 1$. The space between them is empty, it's a hole. We can see that this isn't true for fractions (composed of natural numbers). If we have $a/b < c/d$, we definitely have at least one fraction between them. To see this, take, for example, the sum of the two fractions divided by two. In other words, $(ad + bc)/2bd$. (Do the math: the only thing I'm asking here, and in this whole book, is that you know how to add two fractions.) And you then show that this fractional number is greater than a/b and smaller than c/d, and that it is therefore between them (in fact, it's exactly in the middle). Consequently, the order, on these fractions, is not discrete: it's a dense order, which means, first of all, that between two different fractions there is always at least a third fraction that is different from the first two. But between the first fraction, a/b, and the one that I've just shown is "in the middle" of the space between a/b and $c/d - (ad + bc)/2bd$ – there must therefore exist one more fraction if you perform the same operation. And as this process can continue "to infinity," we'll reach the following very strong conclusion: between any two different fractions there are always an infinite number of other fractions. This gives real meaning to the opposition between discrete order and dense order: where there may be "nothing" (between two successive whole numbers) there is infinity (between two different fractions).

You might ask: why does the proof of infinity, which works for fractions, not work for two successive natural numbers, which are fractions after all? I can write the number n as "n divided by 1," or $n/1$. And the successor of n can be written $(n + 1)/1$. So? So the result of the above calculation is $n + \frac{1}{2}$, which is indeed "between" n and $n + 1$ but has the drawback of . . . not being a natural number. It's possible to calculate it if you're dealing with fractional numbers (positive rational numbers) but not if you're still dealing with natural numbers.

In this way, a structural edifice is gradually built up in which relations seem to prevail over entities, or objects, or even to determine their nature and properties. So it is tempting to reduce all the so-called "intuitive" objects to structural, or formal, manipulations whose principle only obeys the mathematician's decisions or choices. What then "exists" are structured domains, which are accountable only to the formalism by which they are exhibited.

But come on, though! Don't the logical rules for deriving the consequences of the axioms have the status of universal truth? Some mathematicians have invented logics other than traditional binary logic – fine. But those who set out the principles of a new logic continue to think and express themselves in accordance with the principles of identity and non-contradiction of good old traditional logic: they don't say "black" and "white" at the same time, and the rules they lay down are themselves logically consistent in the classical sense of the term. In other words, above and beyond the formal constructions that modern mathematics has generated, isn't there nonetheless a primacy of classical logic, which remains unsurpassable simply because it reflects the a priori *laws of our mind, as Kant claimed?*

Well, you see, the crux of classical logic, what seems to impose it universally on people's minds, is essentially negation. Ever since Aristotle, classical logic has been governed by two main principles. First, the principle of non-contradiction, which I mentioned a little while ago: you can't admit a proposition p and its contradiction, not-p, in one and the same formal system. And second, the principle of the excluded middle: if not-p is false,

p has to be true, and you conclude that p is true. As a result of these two principles, double negation, that is, not-not-p, is equivalent to simple affirmation, that is, p. However, this set-up is being challenged today by the emergence of at least two competing logics, which are relevant in the general field of demonstrative thought.

First, at the beginning of the last century, intuitionist logic rejected the principle of the excluded middle and built consistent formal systems that do without it. It is a logic that's closer to our concrete experience than classical logic. For instance, we all know that in a political meeting there can be not just two mutually exclusive positions but a third position that is ultimately the right one, the one that's really appropriate to the situation. In this case, Position 2 is the negation of Position 1, and this negation obeys the principle of non-contradiction: it is impossible for Position 1 and Position 2, which explicitly contradict each other, to be true at the same time. However, neither one of them is true, since Position 3 is. In such systems, it is generally not the case that the negation of negation is equivalent to simple affirmation.

More recently, paraconsistent logic has emerged. In this type of logical system, it is the principle of non-contradiction that has no general value, whereas the principle of the excluded middle may still be valid. We then get some complicated situations. Take, for example, the case of two people who love the same work of art passionately and, to support their conclusion of admiration, give opposite reasons for it. These reasons may both be true, since there can be a virtually infinite number of interpretations of a work of art. On the other hand, the positivity of the contradiction operates within a first opinion (the two people love the same work of art) to which the excluded middle may apply: between "loving the work" and "not loving the work" there may very well be no third position.

Now, as it turned out, these three logical styles are useful, or even necessary, in certain branches of mathematics. To be sure, mainstream mathematics always operates within classical logic. But in the context of so-called Category theory, which is roughly the theory of relations "in general," with no pre-specification of given objects, paraconsistent logic is clearly operative. In certain categories similar to set mathematics, such as Topoi theory (a topos is a category in which can be defined a relation similar to the classical relation of belonging, the famous ∈), the logic is essentially intuitionistic. Finally, the logical context has in its turn become variable and no longer imposes immutable laws on the mind, even in mathematics. Philosophy has known this for quite some time: in the Hegelian system, the negation of negation is not at all identical to the original affirmation. Its logic is therefore non-classical. In my own system, the logic of pure being, of being *qua* being, is classical, the logic of appearing is intuitionistic, and the logic of the event and of the truths dependent on it is paraconsistent in terms of the Subject.

Let's go back to the original choice, then: which of these two great conceptions of mathematics, the realist one or the formalist one, would you, Alain Badiou, be in favor of?

Between these two visions, and without dwelling any further on the arguments in favor of one or the other of them, I'd choose the former: there is a real "content" to mathematical thought. It's neither a language game – even if complex formalisms are required – nor is it an offshoot of pure logic. I agree with the majority of mathematicians about this issue. Obviously, it's a bit demagogic on my part to use that argument: as you know, even in politics, the concept of "majority" is really not my thing. But still, the truth is that the majority of mathematicians are "Platonists." They don't believe in the second thesis, that of the language game, of total formalism, which is in fact a thesis of essentially philosophical origin. They believe that mathematizable objects or structures "exist" in a certain way. Why do they believe this? No doubt because they've experienced all too often that "something" resists when you practice mathematics, that you come up against a difficult, unyielding reality. But what is it that resists, then? If it's just a matter of a game that has been completely coded through and through, it ought to be like openings in chess, or some such thing. If you know them well, even where much later developments are concerned, you already have a strong strategic superiority. However, generally speaking, mathematicians don't have that impression. On the contrary, they have the impression that the path to the solution of a problem (a path it can occasionally take a few centuries to get to the end of, like Fermat's last theorem, which was no small feat) is a path that makes you touch a real and has a sort of intrinsic complexity. What the exact nature of that real is, is a different discussion. But at any rate you have the feeling of touching an external reality, in the sense that it's not just a fabrication of the mind. If it weren't for that, it would be hard to understand the enormous difficulty and the extraordinary resistance you come up against even in attempting to prove certain properties that really seem to be basic. Take an extremely simple question: twin primes, that is, primes that follow each other, in that the second number is equal to the first one increased by 2. For example, 5 and 7, or 11 and 13, or 71 and 73, and so on. The question is: is there an infinite number of twin primes? Clearly, the farther you go in the sequence of numbers, the "scarcer" they become. But ultimately, through the use of extremely powerful computers, some really big ones were found: twin primes requiring more than 200,000 figures to write out! Nevertheless, compared with the infinity of numbers, even enormous numbers like those are still not much. This is just an illustration that can touch the real of the problem. So? Well, we still don't know whether by continuing the sequence of whole numbers we would still keep finding, "ad infinitum," new twin primes. How is it possible to think that there's no real here other than our own playful invention? How can we not be convinced that the infinity of natural numbers "exists," in a sense that would need to be clarified?

My own, strictly philosophical, conclusion is that, in reality, mathematics is simply the science of being *qua* being, i.e., what philosophers traditionally call ontology. Mathematics is the science of everything that is, grasped at its absolutely formal level, and that's why paradoxical inventions of mathematics may be used in physical investigation. There are some very instructive examples in this respect, the most spectacular among them being complex numbers, the imaginaries, which were invented as a pure game – they were even called "imaginaries" to make it clear that they didn't exist. They could be played with even though they didn't exist. Later, they became a basic tool used in electromagnetism in the nineteenth century, something that no one could have foreseen. Experiences like these keep us from thinking that mathematics is purely and simply a formal, arbitrary game. If, as regards what is, you want to know what it means to think only its being (i.e., not the fact that it's a tree, a pond, a man, but the fact that it *is*), the only way to do so is obviously to think purely formal structures, that is to say, structures indeterminate as to their physical characteristics. And the science of these structures indeterminate as to their physical characteristics is mathematics. It is even mathematics that invented forms like imaginary numbers before it was known, and even before it could be imagined, that they were in fact actualized or actualizable somewhere.

Another very famous, spectacular example is the theory of conics. The definition of ellipses, and the study of them, was introduced in late Antiquity with Apollonius of Perga's *Treatise on Conics*. But it wasn't until the early seventeenth century, or around 2,000 years later, that scientists realized, thanks to Kepler, that the orbit of the planets was an elliptical path, which up until then had been thought of as a circle. In this case, mathematics was clearly the anticipated invention, with respect to pure being, of a number of formal mechanisms that would later, in line with the complex, haphazard development of the natural sciences, be actualized in relevant physical models. This, too, is proof, in my opinion, that mathematics touches a real but in a way that is not experimental, since it is presupposed in experience. It's very clear that Apollonius of Perga thought the being *qua* being of a planet's orbit but without knowing at the time that that's what it was. This is why I reject the theory that mathematics derives from sensory experience. It's the other way around: the real of sensory experience is thinkable only because mathematical formalism thinks, "ahead of time," the possible forms of everything that is. As Bachelard said, even the great instruments that are used in experiments, from telescopes to giant particle accelerators, are "materialized theory,"[4] and presuppose, even in the way they're constructed, extremely complex mathematical formalisms. That, in my opinion, is what solves the mystery of the relationship between the formal sciences such as mathematics and the experimental sciences such as physics.

But does that really suffice to explain the correspondence between the physical laws that govern the real

4 Gaston Bachelard, *Le Nouvel esprit scientifique* (Paris: Presses Universitaires de France, 1934), 14.

and mathematical structures that remain idealities?
Couldn't mathematics exist without matter and the
real obeying the laws of physics, obeying regularities,
which are moreover expressible in mathematical
language?

I'm not claiming that mathematics "needs" the
structural forms it studies to be validated by
experience someday or other. My thesis is: mathematics is ontology, i.e., the independent study
of the possible forms of the multiple as such, of
any multiple, and therefore of everything that
is – because everything that is, is in any case a
multiplicity. This ontology can be developed for
its own sake. The theory of second-degree curves
was invented long before it was applied to the
planets, and the binary number system (using
only 0 and 1) was known before it became the
key to computer coding, and so on. This was so
because the "idealities" you mentioned are actually possible forms of what is, insofar as it is,
and don't need to be experienced as pure forms
to be known, that is, thought, by mathematicians. That said, there can be an inspiration of the
opposite sort. The clearest case is differential calculus. There's no question that its development,

by Leibniz and especially Newton, was to a large extent dictated by the question of movement, by mechanics, itself set in motion by the revolution in astronomy – Kepler, Galileo – and therefore, lying behind it, real observations. It could be said that, in order for the ontological substructure of rational mechanics to be thought, for questions like "What exactly is a body in motion?" or, in particular, "What is acceleration?" to be answered, a veritable mathematical continent had to be opened up, where "smallest difference," "infinitesimal," "derivative of a function at a point," and, finally, "limit," "integral," "differential equation," and so forth, would be spoken about. But as soon as that continent took its purely mathematical form, it developed according to the laws of ontology, which are axiomatic and demonstrative but in no way experimental. You just have to look at Cauchy's final definition of limit. The "intuitive" idea is that of a variable that approaches a point, which is the limit of its movement. This becomes, in ontological, that is to say, mathematical, jargon: "Let S_n be a sequence of real numbers, with n ranging from 0 to infinity. The number L is said to be the limit of this sequence if, for any given real number ε, however small, there exists a number n such that $|L - S_n| < \varepsilon$." This definition makes the supposed – and originally operative – intuition disappear in the icy waters of symbolic calculation.

If the laws of physics happen to obey regularities that can only be formalized in the language of mathematics, it's only because the aim of that language has always been to think the possible forms of everything that's based in its being on some consistency. Now, what exists is in fact made up of multiplicities that have a certain consistency. If this weren't so, it would mean that there would only be totally unpredictable chaos at every moment. As regards this point, experience – unavoidable when it comes to physics – reasonably shows that that this is generally not the case: we observe regularities, consistent objects, a fixed sky, unchanging motions, etc. Hence the intersection of physics and mathematics, which doesn't preclude but rather presupposes the independence of mathematics as an apparatus of thought.

IV

An Attempt at a Mathematics-based Metaphysics

I'd like for us to talk specifically about the way mathematics has inspired your own work in philosophy. The metaphysics you've developed is, if not propaganda (!) for, at least an attempt at, re-entwining philosophy and mathematics. How are they connected to each other in your philosophical system?

What has my philosophical strategy been for about thirty years now? It's been to prove what I call *the immanence of truths*. As I already mentioned, I call "truths" (always in the plural; there's no such thing as "the truth") particular creations with universal value: works of art, scientific theories, politics of emancipation, passionate loves. In a nutshell, let's say: scientific theories are truths about being itself (mathematics) or the "natural" laws of the worlds about which we can have experiential knowledge (physics and biology). Political truths concern the organization of societies, the laws of collective life and its reorganization, all in the light of universal principles such as freedom and, today, primarily, equality. Artistic truths have to do with the formal consistency of finished works that sublimate what our senses perceive: music in terms of hearing, painting and sculpture in terms of sight, poetry in terms of speech . . . Last but not least, the truths of love relate to the dialectical power contained in the experiencing of the world not from the point of view of the One, of individual singularity, but from the point of view of the Two, and hence with a radical acceptance of the other person. These truths, as is clear, are not philosophical in origin or nature. But my aim is to salvage the (philosophical) category of truth that distinguishes between them and names them, by legitimizing the fact that a truth can be:

— absolute, while at the same time being a localized construction;

— eternal, while at the same time resulting from a process that begins in a determinate world (in the form of an event in that world) and thus belongs to the time of that world.

These two properties require truths – whether scientific, esthetic, political, or existential – to be infinite, without resorting to the idea of a God, whatever its form. So I obviously have to begin with the question: on what ontology of infinite-being, which is in no way religious and excludes any transcendence, can I ground my project? It is here that begins the long march in which radical new ideas – especially mathematical ones – concerning infinity, or, more precisely, infinities, come into the picture.

And is that where mathematics becomes necessary?

Broadly speaking, what mathematics ultimately makes possible, how it offers itself – without its knowing or even caring about it – as a speculative resource to philosophers who want to go beyond contemporary relativism and restore the universal value of truths, is what I'd call the possibility of an absolute ontology. Today, it is pretty much accepted, for example, that artistic taste is a question of local culture, of a particular "civilization," or that love is a contingent, terminable choice, which is supposed to provide a contract with mutual benefits for the couple. In politics, it's taken for granted that there is no truth, only volatile opinions that should be formed empirically as much as possible. I, on the contrary, am convinced that there are absolute truths, which, although extracted at the time of their creation from a particular soil (a moment in history, a country, a language, and so on), are nevertheless constructed in such a way that their value becomes universalized. To prove this, I have to show that, within the framework of my ontology of the multiple, a whole new dialectic of the finite and the infinite, and therefore a completely new relationship, too, between our "ordinary" existence and our existence in relation to an absolute truth, can be established. This is what I've also called "living under the authority of an Idea." Or "the true life."

But what is meant by "an absolute ontology"?

What I mean by "an absolute ontology" is the existence of a universe of reference, a site for the thinking of being *qua* being, having four characteristics:

1. It is motionless, or changeless, in the sense that, although it makes possible the thinking of movement, or change, as, for that matter, any rational thought, it is itself foreign to that category.

Consider, for example, the case of movement, as a matter of fact: a real motion is located in a world; it is particular. But the mathematical equation that formalizes the thinking of movement has no specific location itself, except, in fact, its mathematical absoluteness. A stone falls somewhere, but the value of the acceleration of its falling motion, as calculated by post-Newtonian physics, is no different in kind from when it's a question of a different stone, somewhere else.

2. It is completely intelligible in its being on the basis of nothing. Or: there is no entity of which it would be the composition. Or again: it is non-atomic.

Take a revolutionary movement, an uprising that will become historic, such as the storming of the Bastille, let's say. Considered in terms of its pure political value, as a symbol, a reference point, an absolute beginning of a process, this event cannot be broken down into separate units. It's not the result of an addition of factors; it's "absolute" in the sense that, albeit particular in all its components (the people who are there, the series of things that happen, and so on), this particularity disappears in an evental synthesis that can't be broken down into minimal components.

3. So it can only be described, or thought, by means of axioms, or principles, to which it corresponds. There can be no experience or construction of it that depends on an experience. It is radically non-empirical. You could also say that it exists (for thought) even though it *is not*.

This characteristic helps us understand what happens when an event or a work (May '68, Relativity, Héloïse and Abélard, or Picasso's *Guernica*) is said to be an achievement for all of humanity: we then share, in connection with whatever's being talked about, the principles – whether political, scientific, artistic, or amorous – that make it possible to affirm a universal value. Here, description alone doesn't allow us to reach a conclusion. What's needed is the mediation of what constitutes, axiomatically, a principle. All absoluteness is axiomatic and therefore so is any affirmation of the universal value of a work or an event.

4. It obeys a principle of maximality in the following sense: any intellectual entity whose

existence can be inferred without contradiction from the axioms that prescribe it exists by this very fact.

With regard to an ongoing political action, you can speak about the 1917 Russian Revolution, in the sense that you claim allegiance to it, provided you're able to show how a given aspect of your action is consistent with the principles in whose name you regard the Russian Revolution as having an absolute value. In this sense, you exist "timelessly," so to speak, with the Russian Revolution as a co-consequence of these principles.

So we need to renounce God without forfeiting any of the benefits he provides. We must find an immanent and absolute ontological guarantee, which has been completely transferred over to the simple multiple as such, of immanence to the existing world, while still preserving the four key principles of changelessness, composition on the basis of nothing, the purely axiomatic disposition, and the principle of maximality.

That seems like a well-nigh impossible task: in the metaphysical tradition, the guarantee of both infinity and absoluteness is transcendental. Even for Hegel, the Absolute, which is historical, which is

"the becoming of its own self," remains at least One:
it has an infinite unity such that it can still be called
God. You, however, seem to want to absolutize the
multiple as such. Is it there that mathematics comes
to your aid?

That's exactly right. Set theory, which can also
take in all mathematics, as the Zermelo-Fraenkel
formalizations and the French Bourbaki group's
enormous efforts have shown, is an absolute
theory of the undifferentiated multiple (which
originally has no property other than being multiple). Right from *Being and Event* (first published
in 1988), I thus proposed, in order to reach the
goal, to preserve the absoluteness of truths without having recourse to any God, and to simply
incorporate set theory, as a founding mathematical condition, into philosophical reflection.

So was it your sole guide? Mathematics, as the
Ariadne of the philosophical Theseus in the labyrinth
of the Absolute?

At any rate, it can be proved without too much
difficulty that set theory obeys the four principles
of absoluteness that I just mentioned.

Changelessness: This theory is concerned with sets for which the concept of change is meaningless. These sets are extensional, which means they're entirely defined by their elements, by what belongs to them. Two sets that do not have the exact same elements are absolutely different. And so a set as such cannot change, since, just by changing a single point of its being, it loses it altogether.

Composition on the basis of nothing: The theory does not initially introduce any primordial element, atom, or positive singularity. The whole hierarchy of multiples is built upon nothing, in that it only needs the existence of an empty set to be postulated, a set that contains no elements and is for that very reason the name of pure indeterminacy.

Axiomatic prescription: The existence of a given set is only inferred at first either from the void as originally postulated or from the constructions allowed by the axioms. And the guarantee of this existence is nothing but the principle of non-contradiction applied to the consequences of the axioms. Obviously, whether these axioms, historically selected by the mathematical community, are the best ones, or especially whether they are sufficient for thinking multiple-being *qua* being is a question that has no answer *a priori*. It is the history of mathematical and philosophical ontology that will decide. All we can do is admit a principle of openness, which is formulated as the fourth point.

Maximality: An axiom prescribing the existence of a given set can always be added to the theory's axioms, provided it can be proved, if at all possible, that this addition introduces no logical inconsistency into the overall construction. These additional axioms are usually called "axioms of infinity" because they define and postulate the existence of a whole hierarchy of ever more powerful infinities.

This last point is clearly of the utmost importance to my objective of proving the infinity of any truth. The fact that this theory is not and cannot be a monotheistic theory derives from a famous proof: the proof of the non-existence of the One. If indeed we conceive of the One – and this is unavoidable when it comes to an ontological guarantee – as verifying Proposition XV of Book I of Spinoza's *Ethics*: "Whatever is, is in God, and nothing can be or be conceived without God," it must be admitted that any particular multiplicity, any set, is an element of this One, which thus deserves to be called God. And this is what is mathematically impossible: you in fact prove – a really nice, simple proof – that a set of all sets cannot exist. But in that case it's impossible, if the axiomatized multiple is the immanent form of being *qua* being, for *a* being such that all being is in it to exist, since it would have to be a multiple of all multiples, which is a contradiction in terms.

But if the multiples formalized by mathematics do not themselves form a set that is really One, what is the domain of existence of the objects (the multiples) studied by set theory?

The solution is to speak about nothing but the system of axioms at the outset. We'll conventionally call V, the letter V – which can be said to formalize the Vacuum, the great void – the (truly inconsistent, since non-multiple) site of everything that can be constructed from the axioms. What is metaphorically "in V" is what can satisfy the axiomatic injunction of set theory. This means that V is actually just the set of propositions that can be proved from the axioms of the theory. It is a being of language, exclusively. It is customary to call such beings of language "classes." We shall therefore say that V is the class of sets, but bear in mind that this is a theoretical entity that is unrepresentable, or that has no reference, since it is in fact the site of the absolute reference. V exists as the possible and ultimate site of experiments of mathematical thought, of decisions and proofs. But as a set, as a totality, it has no being, precisely because to have a being is to be a multiplicity, and therefore to belong to V, which V itself wouldn't be able to do.

It's with respect to the assumption that such a V "exists," without, however, being, that the question arises of the relations and non-relations between the finite and the infinite, and therefore the rational framework of both an ontology of infinity (or, more precisely, of infinities) and a critique of finitude.

And is this where you got into the intimate connection between mathematical ontology and the philosophical theory of the concept of truth?

Exactly. I simply said this: being is multiplicity. The rational theory of the different possible forms of the multiple is set theory. A truth, like everything that exists, is also a multiple. But how can a multiple bear or be a vehicle for a universal value? I then looked for a clue to this in mathematics. It was an adjective – found in a very contemporary area (it began in 1962) of set theory – that caught my attention: the adjective "generic." There are such things as "generic" multiplicities, defined by the mathematician Paul Cohen. I'm not going to explain what they are; it would be too long and complicated. But I did so meticulously in *Being and Event*. I can nevertheless point out here that Marx, in the *Manuscripts of 1844*, in fact speaks about the proletariat as a "generic" social set. And what does he mean by that? He means that there is a universal truth to the proletariat, that the proletarian revolution will emancipate humanity as a whole. So I was able to introduce the following hypothesis: the being of a truth, what gives it a universal form, is to be a generic set. The "welding together" of a mathematical discovery (Cohen, 1962) and a philosophical proposition (Badiou, 1988) finds a sort of pure form here.

V

Does Mathematics Bring Happiness?

You make an initially rather surprising claim, namely that mathematics, far from being an austere practice reserved for a little caste of specialists, is the shortest way to what you call "the true life," in other words, the happy life. Do you think mathematicians seem happier than other people?

Look, that's no concern of mine! It's no concern of mine because it's uncertain whether creative mathematicians make the best use of mathematics when it comes to existence, to life. The mathematician is totally immanent to mathematical production, in his very definition, and, like all intense subjectivation, that may well involve a good deal of anxiety. Just think, for example, of how violently Grothendieck, who was probably the greatest mathematician of the second half of the twentieth century, broke with the mathematical community and, in a way, with mathematics itself, at least publicly. He took off for the South to raise sheep and devote himself to the environment. That said, this anxiety, about mathematical production, about the close relationship with ontology, also involves moments of elation or ecstasy. And that dialectic exists on a case-by-case basis; I obviously can't propose a theory of it.

But could you perhaps give a personal example?

It's true that we need to have a picture of what mathematical work is in practice, even just in terms of learning how to do it. For instance, I remember one of the nights I spent a long time ago trying to understand the proof of a philosophically fascinating theorem, one of Cantor's fundamental theorems, which essentially says that there are always more subsets of a set than there are elements. I'd like to give you an idea of this night-time experience, of the intense happiness I felt when I understood both the proof and its philosophical implications.

Let's start with what's easiest. A multiplicity, let's say S, is composed of elements, let's say x, y, etc. Note that x, y, and the rest are also sets, but, here, they feature as elements of another set, S.

Any grouping of the elements of S constitutes a subset of S. For example, the pair x and y, which is written as $\{x, y\}$, is a subset of S.

It is certain that there are at least as many subsets of S as there are elements. Indeed, for every element x there exists a subset that is the set of which x is the only element, a set written as $\{x\}$ and called the singleton of x. It's important to understand the difference between x and $\{x\}$: like everything that exists in set theory, x is, as I said, a set, which can contain a large quantity of elements, whereas the singleton of x is a set that contains one and only one element, namely, x.

Since you can make the subset $\{x\}$ correspond to any element x of S, there are definitely as many subsets as there are elements. In other words, there cannot be fewer subsets than there are elements. Now, can there be exactly as many subsets as elements? If that's not the case, then we'll be sure that there are more subsets than elements, since there cannot be either fewer of them or as many . . .

Cantor's theorem doesn't prove directly that there are more subsets than elements but that it's impossible for there to be as many subsets as elements. This is what could be called "indirect reasoning": rather than directly establish the fact that there are more subsets than elements, you arrive at it negatively, via the proof that there can't be as many (and knowing that there can't be fewer).

Negation will play an even bigger part in all this, something that always fascinates me. Once again, we find reasoning by the absurd, which I mentioned in connection with Parmenides and the Greek origin of mathematics. It won't be shown directly that it's impossible for there to be as many subsets as elements; instead, it will be shown that *it is impossible for that to be possible*. It will be assumed that there is a set *S* such that it contains as many subsets as elements. And an "impossible," self-contradictory subset is then constructed, which wrecks the original hypothesis. It is here, in my opinion, that we find the most typical process of mathematical reasoning, as I already said: you assume the false, and through the inadmissible consequences of the false you're compelled to affirm the true.

So let's assume that there exists S with as many elements as subsets. This amounts to saying that there's an exact and complete correspondence between all the elements – x, y, z, etc. – of S and all the subsets (let's call them A, B, C, and so on) of S. What's very striking is that every subset has a name, which is the element corresponding to it; that every element is the name of a subset; that two different subsets have two different element-names; and for two different element-names there exist two different subsets. With these rules (mathematicians call this a "biu-nivocal correspondence" between the elements and the subsets), it can be said that subset A is "named" by an element x, subset B by an element y, and so on. And since the correspondence is total and complete, all the subsets and all the elements are used in this "naming."

It was then that, by what seemed almost like a magic trick to me during the night I'm talking about, an "impossible" subset was constructed. To that end (and this was the brilliant idea), two kinds of elements of S were distinguished: the elements that are in the subset that they name (let's say z is the element of S that names subset B, and it is in subset B) and the elements that are not in the subset that they name (let's say y is the element of S that names subset C, but it is not an element of C). This is a strict and total division: clearly, an element is either in or not in the subset it names; there is no third possibility.

Now let's consider all the elements of S that have the following property: they are not elements of the subset that they name. They do form a subset of S (a subset of S is any set of elements of S). Let's call this subset P (for "paradoxical"). Since it's a subset of S, it is named by an element of S, let's say x_p. There are two possibilities. One is that x_p is not an element of P. In that case, it has the property of the elements that make up subset P, namely, not being in the subset that they name. And therefore it is in P. A blatant contradiction: the consequence of the hypothesis that x_p is not in P is that it is in P! Therefore, it is in P. But, in that case, it has to have the property of the elements that are in P, namely, not being in the subset that they name. But x_p does in fact name P. Therefore, it shouldn't be in P. Another contradiction: the consequence of the hypothesis that x_p is in P is that it isn't in it!

What can we conclude from all this? Clearly, that our original hypothesis (there are as many elements as there are subsets, every element names a subset, etc.) is false. Therefore, there are more subsets than elements.

I eventually arrived at a philosophical thinking of this remarkable process. Within the framework of *reductio ad absurdum*, you strategically assume what you actually think is false. You examine the consequences of such an assumption. And if you're right (that is, if your strategy is to assume the false), you have a chance of finding a truly impossible consequence.

In other words, you reach the true by making the impossible emerge from the false. Well, then, when you've really understood this, in the middle of the night, and you're young and want to be surprised as well as satisfied, you're happy! As an added bonus, you've got a political schema: the fact that there are more subsets than elements in any set means that the richness, the deep resource, of collectivity (the subsets) prevails over that of individuals. At an abstract level, Cantor's theorem refutes the contemporary reign of individualism.

You mentioned a magic trick: using the false to obtain the true via the impossible is actually pretty mysterious.

You could say this: mathematics is wrapped in a sort of mystery, but it's ultimately a mystery in broad daylight. So it is true that, already at this purely practical level, there's the experience of a strange pleasure. Let's indulge in a bit of elementary Freudianism: what we've got here is the childlike mixture of mystery and pleasure, because we'll "see" something we've never seen before. The false will turn into the true. The real will be revealed when an "impossible" object is found. Where Freud is concerned, we know full well what the object is. Where the mathematician is concerned, it's probably not exactly the same thing, but there's a connection, because the mathematical proof is the process of a seeing [*un voir*]. You go back over everything once you've understood it all. But it's no longer the difficult steps, the interminable calculations you get bogged down in, that will fix it in your memory. What will fix it in your memory is your having understood. Now, if you've understood and grasped something, it's because you've seen something you'd never seen before, and it is this ineffable pleasure that will remain.

I think this sensation is paradigmatic of what philosophers call happiness, and it's not something *I* invented either. As you know, at the end of his *Ethics*, Spinoza speaks about intellectual beatitude, intellectual beatitude that is nothing but the fact that one has arrived at an "adequate idea." And the only examples of adequate ideas he gives are in fact connected to mathematics. What he explains is that, with an adequate idea, an idea of the third kind, you're no longer involved in the unfolding of the proof (that would still be the second kind of knowledge), you're no longer involved in the tedium of the proof, in the mathematical exercise, but in its recapitulative synthesis. This is what I call the moment when you've understood – indeed, Lacan, as a true Freudian, speaks about "the moment of understanding." Sure, you've had to go through the tedious steps of the proof, but there's a moment when the light dawns. And that's what Spinoza called the adequate idea, knowledge of the third kind. And it's simply the image of happiness for him, which he calls *beatitudo intellectualis*, intellectual happiness.

But is this happiness of understanding really specific to mathematics? Isn't it experienced in philosophy, too, for example, when our reading of a classical writer suddenly seems to shed new light on our lived experience? And the feeling that you mentioned of having surmounted a problem – isn't that comparable to an athlete's when, after long hours of training, he or she finally succeeds in performing a very difficult movement as though it were second nature?

I'm not about to argue that mathematics has a monopoly on happiness! Nevertheless, the athlete's joy is narcissistic: he or she has succeeded, as an individual self, in doing something. Whereas the joy you feel in mathematics is immediately universal: you know that what you're feeling will also be felt by anyone following the reasoning and understanding it as he or she goes along. Happiness, in mathematics more than anywhere else, is the difficult pleasure of the universal. Sure, philosophy also aspires to guide the subject toward this happiness. But, let me remind you, philosophy, in turning toward its conditions, shows, under the generic name "truths," where the sources of happiness lie, more than being one of those sources itself.

Do you, personally speaking, still take pleasure today in practicing mathematics? Does it afford you a happiness comparable to philosophy?

Let me repeat: I don't claim that philosophy as such produces unparalleled happiness, not at all. The real roots of happiness lie in subjective commitment to a truth procedure: the elation felt in the intense moments of collective political engagement, the pleasure afforded by a work of art that particularly moves you, the joy of finally understanding a complex theorem that opens up a whole array of new thoughts, the ecstasies of love, when, as two, you go beyond the closed, purely finite, nature of the perceptions and emotions of an individual. What I'm saying is that philosophy forges a concept of "Truth" appropriate to the new truths of its times and thereby indicates the possible paths to a becoming-subject, paths blocked by the dominant opinions that establish the supremacy of individual pleasures and/or the cult of conformism and obedience. Philosophy is not a happy practice of the existence of a few real truths; it is instead a sort of presentation of the *possibility* of truths, and so it teaches us the possibility of happiness. That's why I call it "the metaphysics of happiness," not "the theory of happiness." It's in this context that I continue to practice mathematics with great pleasure. Especially since mathematical truths play a vital role in the metaphysics that I'm proposing.

I'd like to come back for a moment to the question of happiness, if indeed a clear definition of it can be given. Do you, following in the footsteps of most of the philosophers of Antiquity, think it is necessarily philosophy's horizon?

I do in fact think that philosophy has no other objective than this: to help anyone understand, in the sphere of his or her own life experience, what a happy direction in life is. You could also say: to provide everyone with the certainty that the true life, the life of a Subject freely guided by a true idea, is possible. Yes, I can say that without hesitation. When Plato – my old master – relentlessly insists that the philosopher is happier than the tyrant, what he is trying to tell us is that anyone who participates in a truth procedure, and does so concretely, vitally, really, not abstractly, anyone who has a life devoted to his or her highest capacities, a life of a free Subject, not a passive or empty life, well, he or she is happier than the pleasure-seeker. Because, in Plato, the tyrant is not primarily a political leader; he's someone who can satisfy all his desires – that's how Plato defines him.

And what is this happiness that's greater than the petty pleasures available in stores? Is there a happiness greater than those pleasures? That's the big question of philosophy. Our societies, domesticated by Capital and commodity fetishism, say that there isn't. But philosophy, tenaciously and right from the start, has striven to make us think that there is ultimately a happiness that isn't necessarily at odds with the petty pleasures, that doesn't prohibit them, but is deeper, more intense, more appropriate, in a nutshell, to the desire of a free Subject, a Subject in a positive relationship with a few truths. You could say this: the commercial mindset of short-lived pleasures, of personal well-being, is like a feeble, dispersed light that leaves us in the dark of life, with only a few openings, a few narrow slits through which light projected from the outside passes. What philosophy says is that we can open much larger windows onto this luminous, freer, less profit-driven outside. We can, as Plato's famous allegory has it, get out of the cave.

But what can that possibly have to do with mathematics?

Well, even though it might seem like a paradox or something very strange, mathematics plays a part in this. Sort of like a scale model. It acts as a model in that there's a very clear relationship in mathematics between the difficulty of understanding, the often long, tedious path of thought, and the happiness afforded by the solution. The original lack of understanding could also be seen as the limits of the individuals that we are, whereas the ultimate comprehension is that of the Subject we have become, which is in contact with the universal. This is very clear, it's something you can experience yourself, which directly connects the effort of thought, the focused effort of thought, and the sort of reward that, albeit universal or even absolute, ultimately owes nothing to anyone or anything except your own effort and can be called, as Spinoza called it, "intellectual beatitude." So, clearly, it's only a model. It doesn't amount to saying: "Do mathematics and you'll all be a lot happier than you are with all the ordinary pleasures!" or "Do mathematics night and day and forget about everything else!"

Not at all. It just means that, here, we have a reduced but accurate model of the possible dialectical relationship between the finitude of the individual who works and makes mistakes and the infinity of the Subject who has understood a universal truth.

However, in the introduction to our discussion, you pointed out that in your philosophical system you distinguish four truth procedures, or to put it another way, four ways of living a life guided by the Idea: in addition to mathematics there are art, love, and politics. But these different ways seem to correspond to completely different existential experiences of happiness. In what sense is mathematics the privileged matrix of them all?

Again, I claim that mathematics – aside, of course, from its major philosophical extension, since it is ontology – serves as a model, a reduced-scale model, perhaps, depending on nothing other than the concentration of pure thought. I don't claim that it is in and of itself a sort of ultimate bliss. Obviously, if, as can be shown, the four conditions are spaced out from mathematics to poetry, with the other sciences, politics, love, and

the other arts in between, we could take everything I said and try to see what the difference is
exactly between mathematics and poetry in terms
of the happiness these conditions afford: that
would be another way of going about it. Between
mathematics and poetry there's love and politics,
that is, the minimal form of relationship to the
other, that basic cell of relationship to the other
that is love, and the maximal form, which is the
relationship to humanity as a whole, a point that
should always be the concern of politics but is
only so in genuinely communist politics.

The four conditions are separate from each
other at first. There are of course overlappings
between them, but they each operate on their
own, and they are inscribed in philosophical
reflection in different ways. For example, love is
the existential matrix of the thinking of difference
as such. It's the possibility of living in difference,
not in indifference, i.e., of experiencing how the
world can be approached or dealt with from the
point of view of the Two, not just from the point
of view of the One. And so love is the existential
learning of the dialectic, that is, of the richness of
difference. This is one of the reasons why there
are so many works of literature about the power

of love, precisely to overcome artificial differences and accept going beyond identity. Romeo and Juliet belong to two clans that are normally supposed to remain absolutely separate and hate each other. Romeo and Juliet's love is the relationship they weave in the name of their difference – a difference that will be creative and not absorbed back into criminal hostility. This is why, at the very heart of the impossible and the threat of death, there is the dawn of love of Romeo and Juliet, who express their happiness in tones of rare beauty.

So this doesn't need to be related to mathematics. But it's by no means incompatible with it: if you do mathematics with someone you love, which I've done on a number of occasions in my life, and if you try to find the solution to the same difficult problem together, well, it's an amorous and mathematical experience at one and the same time. When you find the solution to the problem together, your joy is doubled, and you're not sure which of the registers it belongs to.

Specifically in politics, do you think that mathematics can be a valuable requisite?

DOES MATHEMATICS BRING HAPPINESS?

There's no obvious connection between mathematics and politics. The zero degree of connection is the counting of the votes on election night. To be sure, you have to deal with the concepts of absolute majority and qualified majority, the percentage of abstainers, and other such counting of blank ballots, which are different from invalid ballots. But even so, that's still very basic. And since, in my opinion, the specifically political stakes are practically non-existent – except for a few details, the people elected all do the same things – you can't speak of truth or, therefore, of happiness either. There's only the very short-lived pleasure of the person elected and his or her cronies.

In my view, it's the following question that matters: Do you think it's possible, in politics, to reach decisions that really result from rational deliberation? Can there be such a thing? Or are there ultimately only opinions in politics, as Plato, who tried to fight for a politics of truth, thought? I don't think he found the solution, but that was indeed his objective. And knowing what a real argument is, an argument whose conclusion anyone who follows all its steps must agree with – and that's the only way mathematics

reaches an agreement that's in some sense absolute – well, that's important in every area where
deliberation is required. Just to know that there
are methods for reaching strong agreement – at
any rate when the problem has been clearly laid
out and everyone discussing it is really interested
in finding a solution to it – can be useful when
you need to come up with a positive solution col-
lectively in a difficult situation. Of course, that in
no way suffices to define a politics. But it can help
to change the methods of politics, methods that
are often a somewhat murky combination of real
but unclear or poorly explained common interests, imaginary representations, and inadequate
or outmoded symbolism. If we want to avoid
that, we need to have a common standard to discuss the decision to be made. Mathematicians do
in fact have a common standard when they examine a problem, and that's why they can come to
an agreement on the proof, or, if it's false, say
so – and the person who proposed the proof will
have to agree as well.

A rational method of political discussion
remains an ideal, even if everyone who has ever
been an activist knows that there can be thrilling
meetings, particularly in working-class circles,

precisely because the conclusion, the operative, unifying slogan, was the result of a long and very efficient process. And that solution is a real collective joy. At a very general level, the question could be formulated like so: is political discourse forever condemned to being nothing more than rhetoric? Those who think it is, who think that political discourse is a rhetoric of victory, are the sophists. Here are our good old opponents from the fourth century BCE again. The sophists coaxed people into using a rhetoric of victory whatever their personal beliefs may have been and regardless of any "truth" whatsoever.

Unfortunately, rhetoric is today's political language. It is a rhetoric of promises that won't be kept, a rhetoric of the impossible agenda, a rhetoric of bogus necessity. Beneath this rhetoric, a number of decisions are made, in meetings that are usually secret or set up to lead to the desired conclusion, in the service of a number of vested interests whose influence is impossible to counteract. Sometimes the rhetoric even results in a disastrous decision, including for the people who proposed it. Parliamentary politics, falsely called "democratic," is a world controlled by a mix of unclear interests, often vulgar, or even hateful, emotions, false knowledge, and irrational rhetoric.

If I must praise mathematics, including in the field that you're suggesting, I would say this: a sustained and ongoing exercise of true discursive rationality would counteract or mitigate our exposure to seductive rhetorics devoid of real substance. Therefore, I think that, with the help of a totally overhauled education, everyone should acquire, before the age of 20, an extensive knowledge of modern mathematics, enabling them to master the recent advances in this science and to pursue it if they so desire without being held back by ignorance – often attributed, moreover, to the lack of some imaginary knack for the subject – because mathematics offers exercises geared toward developing a discursive rationality that makes it possible for people to agree on difficult decisions.

Actually, mathematics is the best of human inventions for practicing something that's the key to all collective progress and individual happiness: rising above our limits in order to touch, luminously, the universality of the true.

Ultimately, mathematics, in your view, offers us the chance to experience in all its purity and simplicity a subjective relationship to truth. Is that why it's a model of "the true life" in the other areas of life, such as love or politics?

That's exactly right. Mathematics' simplicity, its purity, its lack of compromise with the average state of affairs and the jumble of opinions – all this guides the thinking and the existence that are devoted to it for a time toward "the true life." And just consider the paradox: most people object to mathematics on account of its complexity – as well as its blatant lack of existential meaning. But really! It's mathematics' simplicity, the fact that it is unambiguous, with nothing hidden or obscure about it, with no double meanings or deliberate deception, that can fill us with wonder. And its indifference to dominant opinions is a perfect example of freedom. In that sense, yes, to attain comparable simplicity and universality in politics or love can be accepted as an ideal of life.

Conclusion

Your praise of mathematics has emphasized its importance not only for philosophers but for anyone aspiring to what you call "the true life." So that prompts a final, critical question: how can we get people to discover – or rediscover – mathematics, and, above all, how can we get them to love it?

Well, now you're asking me a question I'm particularly sensitive to. I think that the way mathematics functions in the teaching profession overall isn't what it should be, and may never have been exactly what it might have been. The reason for this is that when you teach mathematics, you first have to convince the students that it's interesting. You shouldn't say: "It's something you've got to know: just learn such and such, and that's it." At most, that lets you deal with the most urgent things first, by teaching the children the multiplication tables, for example. That's only a kind of pragmatic approach to counting. But if it's a matter of true mathematics, the mathematics that exposes you to problems as important as they are complex, you absolutely must instill the feeling that it's interesting, as I already said regarding the transmission of knowledge of any kind.

So how can we stimulate this feeling? It's all about the notion of a problem solved. I'm convinced that a child, even a very young one, can be interested in the idea of solving problems. Because children naturally love riddles. They're curious and love discovering something they've never encountered before. Everything should revolve around this revealing, this mystery solved. Teaching should be completely focused on this objective of producing in children, adolescents, and ultimately everyone, the feeling that what's extraordinary about mathematics is that, in sometimes surprising and unexpected ways, you can solve riddles that are formulated very clearly and precisely but are nonetheless real riddles. When it comes to this issue, you shouldn't hesitate to enter the world of games, because solving a problem is also a feature of games, after all. This doesn't necessarily entail a conception of mathematics as a game, but it does stimulate interest. Moreover, in some magazines and newspapers you can find math puzzles, and I don't think that approach should be scorned, any more than it is sensible to criticize crossword puzzles, which teach spelling and a rather sophisticated semantics.

Along with the methods of convincing learners that mathematics is interesting, there are also two points of support outside mathematics.

First of all, the history of mathematics, which should be presented in a living, breathing way, not by sticking to a boring, systematic exposition of solutions. Don't stick to the solutions, or even primarily to them, but to what's interesting about the problem as a riddle that was finally solved after many trials and tribulations. It's exciting to understand how and why a little Greek theorem was discovered, in what circumstances, what it was used for, what it became thereafter, how philosophers commented on it, and so on. Take the famous example used by Plato in the *Meno*: how to construct a square whose area is double that of a given square. It might be a problem concerning a conflict between farmers, about arable land. In the dialogue, Socrates proposed this problem to a slave boy who happened to be around. And he showed that the slave boy, after a bit of trial and error, could easily understand the proof, which establishes that the square that doubles the area of a square *ABCD* is the square whose side is the diagonal of the first square, say *AC*. This can actually be seen as soon as you make a drawing of it, as soon as you draw the square on the diagonal. But what's behind the slave boy's intuitive understanding of the problem is in reality extremely complex and puzzling. Indeed, as anyone can easily see, the area of a square is the product of two of its sides. Let's say that the length of the sides of the first square *ABCD* is 1 (1 meter, for example). Its area will be 1 x 1, or 1 (square meter). The area of the second square, constructed on diagonal *AC*, will be, as the drawing made of it shows, the double of that, hence 2 (square meters). So what is the length of the side of the second square, diagonal *AC*? The ratio between the two areas is clear: it's 2:1, hence 2. What is the ratio of the two sides? Let's apply the Pythagorean Theorem to the right triangle *ABC*. We've got $AB^2 + BC^2 = AC^2$.

And since $AB = BC = 1$, we've got $1^2 + 1^2 = AC^2$. That is, $1 + 1 = AC^2$, or $2 = AC^2$. So the length of diagonal AC must be a number whose square is equal to 2. Today, this is called "the square root of 2." But the problem is that this number is neither a whole number nor a rational number, i.e., a ratio of two whole numbers, which is also called a fraction. For the Greeks, who, where numbers were concerned, only knew whole numbers and their ratios, the number that measures the length of the diagonal, our modern square root of 2, didn't exist. The trace that remains of this is that, even today, numbers of this type are called "irrational." Thus, the little geometry problem "construct a square whose area is double that of a given square," whose solution is intuitive, opens onto an arithmetical abyss, which would occupy Greek mathematicians for 300 years and would raise very difficult problems concerning the so-called irrational numbers right up to today. That's why the history of problems, the commentary on them, the difficulty in finding the solution to them, is to my mind part of the teaching of mathematics.

The second point of support, in addition to the history of mathematics, is to be armed with philosophy, because, in the final analysis, what's interesting about mathematics is also to wonder what mathematics is. And this question, as we saw, is specifically philosophical; there's no other place where it's explored. That's why I think philosophy should be taught right from pre-school, really. It's well known that three-year-old children are far better metaphysicians than eight-een-year-old ones, because they wonder about all the questions of metaphysics. What's nature? What's death? What's the Other? Why are there are only two sexes and not three? All of that is an established terrain of investigation. Just as I think that a lot of basic mathematics can be learned by the telling of stories and the solving of fun riddles, so, too, I think that the highest philosophy is also involved in all this. It's really a shame that philosophy is only begun with great difficulty in the final year of high school. There were some very vigorous efforts, particularly on the part of my late lamented colleague Jacques Derrida, to get philosophy taught in 9th or 10th grade. We have unfortunately not made the slightest pro-gress. Philosophy is still an endangered discipline in the final years of high school and mathematics a deplorable operator of social selection. Well,

CONCLUSION

I suggest they both be taught in the last year of preschool: five-year-old kids will surely be able to make good use of the metaphysics of infinity and set theory!